LEARNING
MULTIPLICATION
USING
LEGO® BRICKS
STUDENT EDITION

Dr. Shirley Disseler

COMPASS

Learning Multiplication Using LEGO® Bricks — Student Edition

Brigantine Media/Compass Publishing
211 North Avenue
St. Johnsbury, Vermont 05819
Phone: 802-751-8802
Fax: 802-751-8804
E-mail: neil@brigantinemedia.com
Website: www.compasspublishing.org

ORDERING INFORMATION
Quantity sales
Special discounts for schools are available for quantity purchases of physical books and digital downloads. For information, contact Brigantine Media at the address shown above or visit www.compasspublishing.org.

Individual sales
Brigantine Media/Compass Publishing publications are available through most booksellers. They can also be ordered directly from the publisher.
Phone: 802-751-8802 | Fax: 802-751-8804
www.compasspublishing.org
ISBN 978-1-9384065-9-1

CONTENTS

FINDING FACTORS

1. Place a 2x8 LEGO® brick on your base plate. Draw it on the base plate diagram below. How many studs are in your model? _16_

2. Next to the 2x8 brick, add two bricks of the same size that are each ½ of the 2x8 brick. Which bricks did you add? _cX4_____ Draw those bricks below.

3. Find the next set of bricks that you can add to the model that are the same size and are equal in length and width. Which bricks did you choose, and how many did you add? _____ Draw those bricks below.

4. Can you add another set of bricks that are equal in length and width to this model to show another factor? What bricks did you find? _____ Draw those bricks below.

5. List all the factors for the 2x8 brick. _____

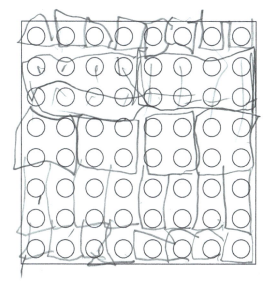

Lucas   Math

6. Build a model to show all the factors of 6. Draw your model and explain your thinking.

The factors of 6 are
1x6   1, 2, 3, 6
2x3
3x2
6x1

7. Build a model to show all the factors of 8. Draw your model and explain your thinking.

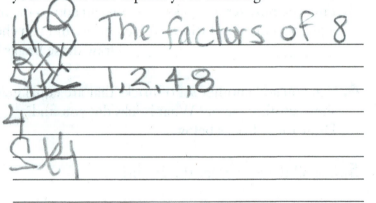

The factors of 8
1, 2, 4, 8

More problems to practice:

8. Build a model to show all the factors for 12. Draw your model and explain your thinking.

9. Build a model to show all the factors of 24. Draw your model and explain your thinking.

_____

_____

_____

_____

_____

_____

_____

_____

_____

_____

_____

_____

_____

Assessment:

1. What is a factor?

_____

_____

_____

2. What word means "the answer to a multiplication problem"? _____

3. Choose a number you have not used in this lesson. Build a model with bricks to show all the factors of this number. Note: you do not have 5, 7, or 9-stud bricks. Draw your model and explain your thinking.

_____

_____

_____

_____

_____

_____

_____

MULTIPLICATION USING SET MODELS

1. Use LEGO® bricks to model 4 sets of 6. Draw your model.

Explain your model.

_____

_____

_____

_____

_____

_____

_____

_____

How many studs are there in all? _____

Write a math sentence for this problem._____

2. Use bricks to model 5 sets of 3. Draw your model.

Explain your model.

_____

_____

_____

_____

_____

_____

_____

_____

_____

_____

_____

_____

_____

_____

How many studs are there in all? _____

Write an equation for this problem. _____

3. Use bricks to model 2 sets of 6.
Draw your solution on the base plate.

Explain your model.

_____

_____

_____

_____

_____

_____

_____

How many studs are there in all? _____

Write an equation for this problem. _____

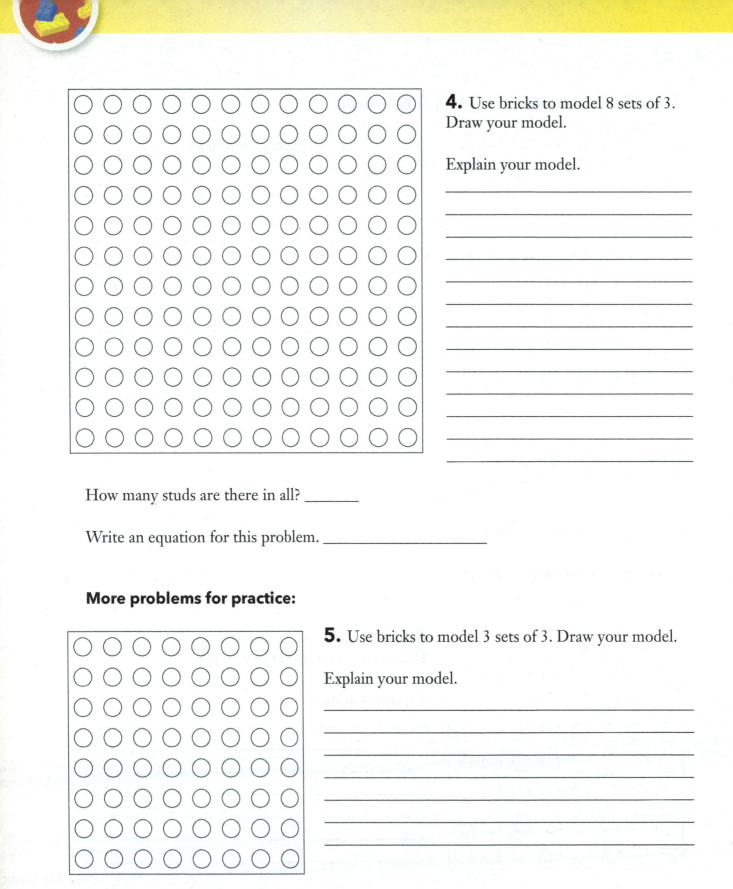

4. Use bricks to model 8 sets of 3. Draw your model.

Explain your model.

_____
_____
_____
_____
_____
_____
_____
_____
_____
_____
_____
_____
_____

How many studs are there in all? _____

Write an equation for this problem. _____

More problems for practice:

5. Use bricks to model 3 sets of 3. Draw your model.

Explain your model.

_____
_____
_____
_____
_____
_____
_____

How many studs are there in all? _____

Write an equation for this problem. _____

6. Use bricks to model 4 sets of 5. Draw your model.

Explain your model.

_____

_____

_____

_____

_____

_____

_____

_____

_____

_____

_____

_____

How many studs are there in all? _____

Write an equation for this problem. _____

Assessment:

1. Identify the factors in the problem: 4 x 3 = 12   _____ and _____

2. Identify the product in the problem: 5 x 1 = 5   _____

3. Explain each of the parts of the problem: 6 x 3 = 18

_____

_____

_____

_____

4. Create a multiplication problem. Write the problem: _____

With bricks, build a model of your problem. Draw your model.

Using your model, explain how you know the solution to your problem.

_____

_____

_____

_____

_____

_____

_____

FACT FAMILIES

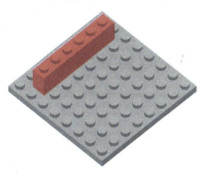

1. How many studs are on this brick? _____

With bricks, model all the fact families for this number. Draw your model and label each fact family with the multiplication sentence.

[grid of 8 × 8 circles]

List all the fact families for 6. _____

_____

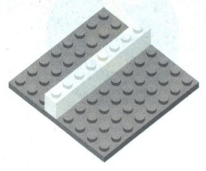

2. How many studs are on this brick? ____

With bricks, model all the fact families for this number. Draw your model and label each fact family with the multiplication sentence.

List all the fact families for 8. _____

_____

3. With bricks, model the fact families for 12. Draw your model.

List all the fact families for 12.

_____
_____
_____
_____
_____
_____
_____
_____
_____
_____

4. With bricks, model the fact families for 16. Draw your model.

List all the fact families for 16.

_____
_____
_____
_____
_____
_____
_____
_____
_____
_____

5. With bricks, model the fact families for 24. Draw your model. Hint: Think about which bricks to use, since there is no brick with exactly 24 studs.

List all the fact families for 24. _____

_____

More problems to practice:

6. With bricks, model the fact families for 9. Draw your model. Hint: Use 9 single bricks for the number nine or a combination of the same color of other bricks to add up to 9. Explain your thinking.

7. With bricks, model the fact families for 36. This will require a combination of bricks. Draw your model and explain your thinking.

_____

_____

_____

Assessment:

1. What is a fact family?

_____

_____

_____

2. Write the fact family for 10.

_____

_____

_____

3. Build the fact family for 18 using bricks. Draw your model and label each part.

BLOCKS AND BRICKS GAME

Play this game in pairs. Each player rolls a number cube to create his/her own multiplication problem to solve.

For each pair of players, you need two number cubes (dice), two large base plates, and LEGO® bricks. At minimum, have these bricks for the game:

Size	Number
1x1	20
1x2	10
1x3	10
1x4	10
1x6	3

You also need some long bricks (1x10 and 1x12, preferably) to build the game boards.

Game Directions:

1. Each player makes this chart:

Multiplcand	Multiplier	Final Model Sketch	Problem/Solution

2. Player 1 rolls one number cube. That roll is the multiplier.

3. Player 1 builds a game board with blocks representing the multiplier. (In this example, the multiplier rolled is 2, so the game board is built with 2 blocks.)

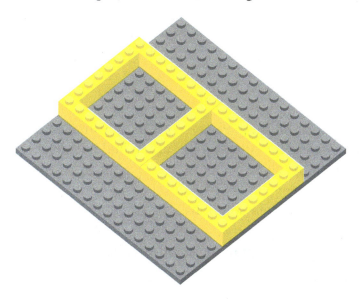

4. Player 1 rolls the number cube again. This roll is for the multiplicand. (In this example, the number rolled is 3.)

5. Player 1 models the multiplicand by placing bricks with that number of studs in each block of the game board. (In this example, Player 1 places three 1x1 bricks in each block.)

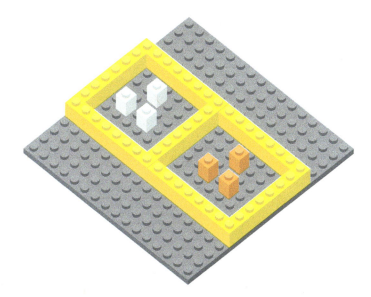

6. Player 1 counts the studs of the bricks in all the blocks to find the solution to the problem.

7. Player 1 records the problem in his/her chart.

Multiplier	Multiplicand	Final Model Sketch	Problem/Solution
2	3		2 x 3 = 6 studs

8. At the same time, Player 2 does the same procedure as Player 1, creating his/her own multiplication problem to solve and modeling the solution. Be sure to build a model for each roll and draw the model in the box in the chart.

9. Both players compare their models and discuss their solutions.

10. After four rounds, add up the total number of studs in the solution column. The player with the highest number wins the game!

MULTIPLICATION USING PLACE VALUE/BUNDLING MODELS

1. What does it mean to multiply 2 x 25?

_____

_____

_____

Use LEGO® bricks to model 2 x 25. Draw your model.

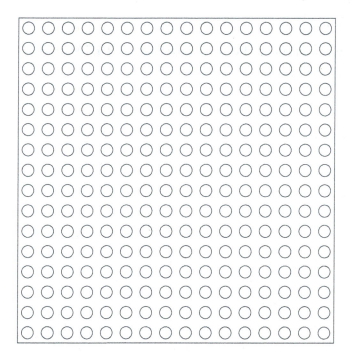

Model 25 x 2 with bricks. Draw your model. Explain why this model looks different from 2 x 25.

_____
_____
_____
_____
_____
_____
_____
_____
_____
_____
_____
_____

On the 2 x 25 model, bundle the tens together. Compose the ones into a set of 10. Draw the solution. What is the product? Explain your thinking.

_____
_____
_____
_____
_____
_____
_____
_____
_____
_____
_____
_____
_____

2. With bricks, make a place value model of 3 x 12.

What does the 3 represent?

_____

What does the 12 represent?

_____

Model 12 x 3 with bricks. What does the 3 represent?

_____

What does the 12 represent?

_____

Draw your models.

Bundle the model of 3 x 12 to show how the 3 sets of 12 come together to form the solution. Draw your final solution model. What is the product? Explain your thinking.

3. Model 3 x 13 with bricks. Bundle the 3 sets of 13. Draw your model. Label your drawing. Show your solution.

How would the model for 13 x 3 look different? Is the solution the same? Explain your thinking.

_____

_____

_____

4. With bricks, build models that show the difference between 2 x 24 and 24 x 2. Draw your models.

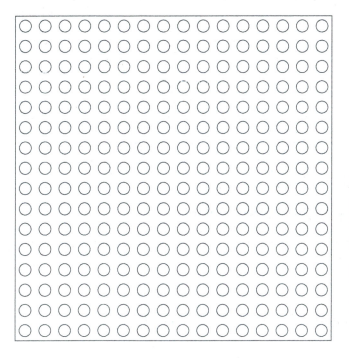 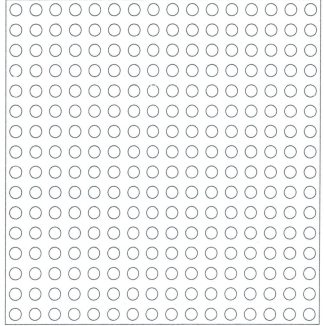

Show the bundling of 2 sets of 24 on the left diagram to get the solution. Show the bundling of 24 sets of 2 on the right diagram to get the solution.

How are these two models alike and different? Explain your thinking.

_____

_____

_____

5. With bricks, build models that show the solution of 3 x 14 using bundling. Draw your models.

Explain all the steps and how you came to the solution.

_____

_____

_____

6. With bricks, build models that show the solution of 4 x 22 using bundling.

Explain all the steps and how you came to the solution.

_____

_____

_____

More problems for practice:

7. Model 2 x 16 with bricks and use the bundling process to solve the problem. Draw your models. Explain your steps and your solutions.

_____

_____

_____

8. Model 2 x 28 with bricks and use the bundling process to solve the problem. Draw your models. Explain your steps and your solutions.

_____

_____

_____

Assessment:

1. What does it mean to _bundle_ in multiplication?

_____

_____

_____

2. How is 23 x 2 different from 2 x 23?

_____

_____

_____

3. Build models to show the similarities and differences between 12 x 2 and 2 x 12. Draw and explain your models.

_____

_____

_____

MULTIPLICATION USING ARRAY/AREA MODELS

1. Make an array/area model of 3 x 2 using only one brick.

Use the same sized brick to make an array/area model of 2 x 3.

Draw your array/area models. Label each model to show which multiplication sentence it represents.

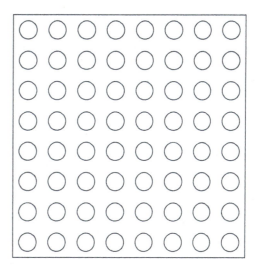

_____   _____

What is the difference between the two models? Explain your thinking. Be sure to include the word _product_ in your explanation.

_____

_____

_____

2. Make an array/area model of 4 x 4 using two bricks. Draw your model.

What do you notice about this model that is different from the 3 x 2 and 2 x 3 models?

_____

_____

_____

What is the multiplication sentence for this model? _____

What is the product? _____

3. Build two more array/area models that have the same rule as the one that models 4 x 4. Draw your models. Label each model with a multiplication sentence containing the factors and the product.

_____   _____

Explain how these two models are similar to the model of 4 x 4.

_____

_____

_____

4. Make two different array/area models using bricks that show 4 x 6 studs and 6 x 4 studs. You will need to use more than one brick. Draw and label your models.

_____   _____

Explain how these two models are alike and how they are different.

_____

_____

_____

What is the product? _____

5. Make two different array/area models using bricks that show 3 x 6 studs and 6 x 3 studs. You will need to use more than one brick. Draw and label your models.

_____   _____

Explain how the two models are alike and how they are different.

_____

_____

_____

What is the product? _____

Assessment:

1. What is an array?

_____

_____

_____

2. How does the orientation of the model change the meaning of the model?

_____

_____

_____

3. With bricks, build an array/area model for 2 x 6 and an array/area model for 6 x 2. Draw your models. Explain how they are alike and different.

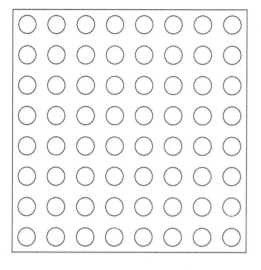 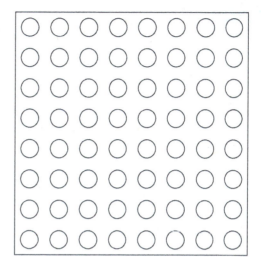

_____

_____

_____

MULTIPLICATION MODELING CHALLENGE

Set modeling:

1. How many studs are in each block? _____

How many blocks are shown? _____

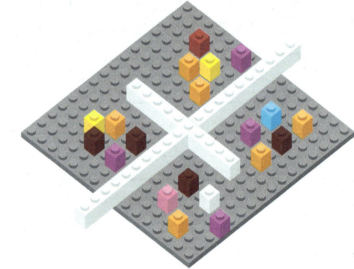

2. How many studs are in each block? _____

How many blocks are shown? _____

3. What is the difference between these two models?

_____

_____

_____

4. Write a multiplication problem for the first model. _____

5. Write a multiplication problem for the second model. _____

Place value modeling:

6. With bricks, build a place value model to show 3 x 14. Draw your model and explain your thinking.

_____

_____

_____

_____

_____

_____

_____

_____

_____

_____

_____

_____

_____

Array/area modeling:

7. With bricks, build an array/area model to show 4 x 8. Draw your model and explain your thinking.

_____

_____

_____

_____

_____

_____

_____

_____

_____

_____

_____

8. With bricks, build an array/area model of a square number. Draw your model and explain your thinking.

_____

_____

_____

_____

_____

_____

_____

_____

_____

_____

_____

More problems for practice:

9. With bricks, build an array/area model of 4 x 6. Draw your model and explain your thinking.

_____

_____

_____

_____

_____

_____

_____

_____

10. With bricks, build a set model of 2 sets of 6. Draw your model and explain your thinking.

_____

_____

_____

_____

_____

_____

_____

_____

11. With bricks, build a place value model of 4 x 12. Draw your model and explain your thinking.

_____

_____

_____

_____

_____

_____

_____

_____

_____

_____

_____

_____

Assessment:

1. What does a set model show?

_____

_____

_____

2. What does a place value model show?

_____

_____

_____

3. Make an array/area model of 2 x 6. Draw and explain your model.

_____

_____

_____

_____

_____

_____

_____

_____

_____

4. Make a place value model of 3 x 16. Draw and explain your model.

_____

_____

_____

_____

_____

_____

_____

_____

_____

_____

_____

_____

_____

MULTIPLYING TWO-DIGIT NUMBERS BY ONE-DIGIT NUMBERS

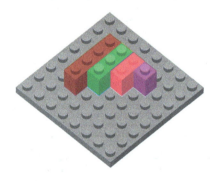

1. What are the place values of the bricks in this model?

The red brick (1x4) is equivalent to _____

The green brick (1x3) is equivalent to _____

The pink brick (1x2) is equivalent to _____

The purple brick (1x1) is equivalent to _____

The whole number in this model is _____

The expanded form of this number is _____

2. Use the expanded form method to model the number 1,234 with bricks. Draw and explain your model. Be sure to explain what each brick represents in the number.

_____

_____

_____

_____

_____

_____

_____

_____

3. Use the expanded form method to model the problem 2 x 10 with bricks. Draw your model. Label the bricks that show the multiplier and the multiplicand.

With bricks, model the process of multiplying two sets of ten. Draw your model and label it to show the sets and the solution. Explain your thinking.

_____
_____
_____
_____
_____
_____
_____
_____
_____
_____

4. With bricks, model 2 x 23. Draw your model on the left base plate diagram. Be sure to show two sets using set markers.

With bricks, model the process of decomposing the sets to find the solution. On the right base plate diagram, draw the process. Label and explain your model. Use the terms *multiplier* and *multiplicand* in your explanation. Write the expanded form of the number.

_____
_____
_____

5. With bricks, model 3 x 34. Draw your model on the left base plate diagram. Be sure to show three sets using set markers.

With bricks, model the process of decomposing the sets to find the solution. On the right base plate diagram, draw the process. Label and explain your model. Use the terms *multiplier* and *multiplicand* in your explanation. Write the expanded form of the number.

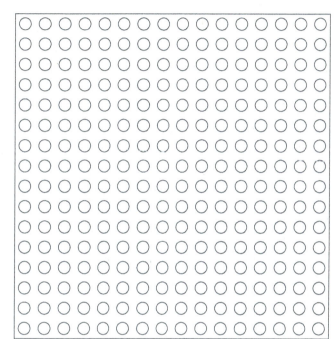

_____

_____

_____

6. With bricks, model 2 x 22. Draw your model on the left base plate diagram. Be sure to show two sets using set markers.

With bricks, model the process of decomposing the sets to find the solution. On the right base plate diagram, draw the process. Label and explain your model. Use the terms *multiplier* and *multiplicand* in your explanation. Write the expanded form of the number.

_____

_____

_____

7. With bricks, model 3 x 32. Draw your model on the left base plate diagram. Be sure to show three sets using set markers.

With bricks, model the process of decomposing the sets to find the solution. On the right base plate diagram, draw the process. Label and explain your model. Use the terms *multiplier* and *multiplicand* in your explanation. Write the expanded form of the number.

_____

_____

_____

8. With bricks, model 4 x 15. Draw your model on the left base plate diagram. Be sure to show four sets using set markers.

With bricks, model the process of decomposing the sets to find the solution. On the right base plate diagram, draw the process. Label and explain your model. Use the terms *multiplier* and *multiplicand* in your explanation. Write the expanded form of the number.

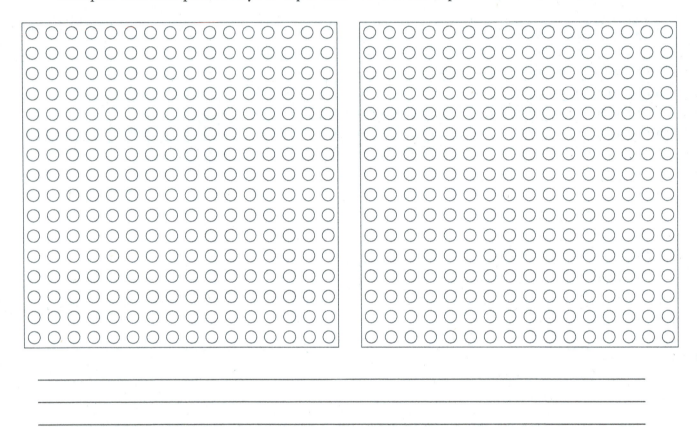

_____

_____

_____

Challenge: Build a model of a multiplication problem and have your partner show the process and explain the problem.

More problems for practice:

9. With bricks, model 3 x 16. Draw your model on the left base plate diagram. Be sure to show three sets using set markers.

With bricks, model the process of decomposing the sets to find the solution. On the right base plate diagram, draw the process. Label and explain your model. Use the terms *multiplier* and *multiplicand* in your explanation. Write the expanded form of the number.

_____

_____

_____

10. With bricks, model 3 x 14. Draw your model on the left base plate diagram. Be sure to show three sets using set markers.

With bricks, model the process of decomposing the sets to find the solution. On the right base plate diagram, draw the process. Label and explain your model. Use the terms *multiplier* and *multiplicand* in your explanation. Write the expanded form of the number.

_____

_____

_____

Assessment:

1. Identify the *multiplier* in these problems:

    a. 5 x 14 _____

    b. 7 x 26 _____

2. Identify the *multiplicand* in these problems:

    a. 2 x 35 _____

    b. 4 x 20 _____

3. With bricks, build a model to show the process of multiplying 3 x 12.

Draw your model and the multiplication process on the base plates. Explain your process. Be sure to include the vocabulary words *multiplier*, *multiplicand*, *place value*, and *sets* in your explanation.

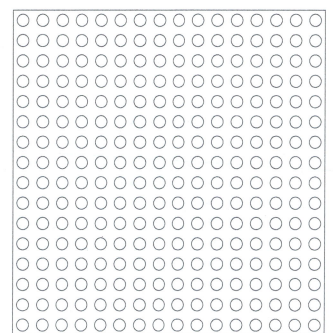

_____

_____

_____

9

MULTIPLYING LARGER NUMBERS

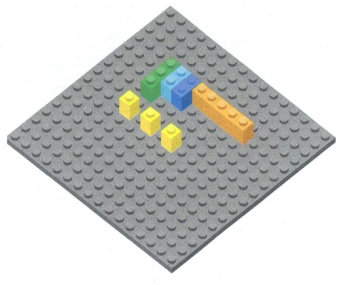

1. What problem does this model show? _____

What is the multiplicand? _____

What is the multiplier? _____

Use bricks to make a model that shows the sets of the multiplicand. Draw your model. Explain your thinking.

Remove the set markers and pull together the solution on your brick model. Draw your model.

What does the solution show?

2. Use bricks to model the problem 3 x 2,232. Draw your model and explain your thinking.

_____
_____
_____
_____
_____
_____
_____
_____
_____
_____
_____
_____
_____

Use bricks to model the sets for the problem. Draw your model and explain your thinking.

_____
_____
_____
_____
_____
_____
_____
_____
_____
_____
_____
_____
_____
_____

Use bricks to model the solution to the problem. Draw your model and explain your thinking. Include the expanded form in your explanation.

_____

_____

_____

3. Use bricks to model the problem 3 x 325. Draw your model.

Use bricks to model the sets for the problem. Draw your model and explain your thinking.

_____
_____
_____
_____
_____
_____
_____
_____
_____
_____
_____
_____

Use bricks to model the solution to the problem. Draw your model and explain your thinking. Include the expanded form of the solution.

_____
_____
_____
_____
_____
_____
_____
_____
_____
_____
_____
_____

4. Use bricks to model the problem 4 x 1,323. Draw your model.

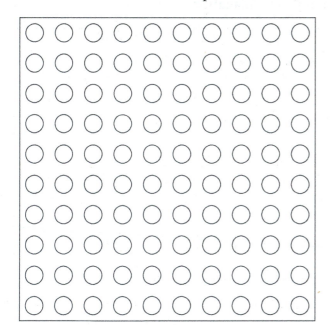

Use bricks to model the sets for the problem. Draw your model and explain your thinking.

_____
_____
_____
_____
_____
_____
_____
_____
_____
_____
_____
_____
_____

Use bricks to model the solution to the problem. Draw your model and explain your thinking. Include the expanded form of the solution in your explanation.

_____
_____
_____

5. Use bricks to model the problem 5 x 2,241. Draw your model.

Use bricks to model the sets for the problem. Draw your model and explain your thinking.

_____

_____

_____

Use bricks to model the solution to the problem. Draw your model and explain your thinking. Include the expanded form of the solution in your explanation.

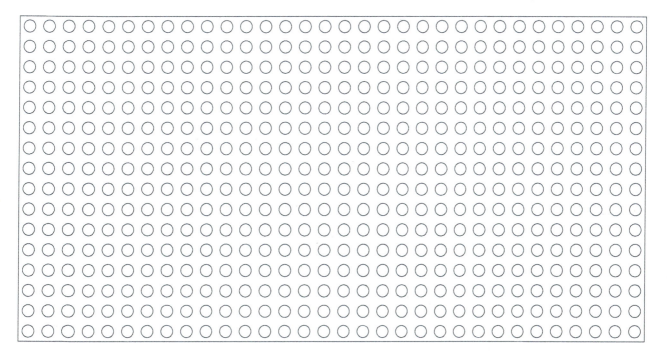

_____

_____

_____

More problems for practice:

6. Use bricks to model the problem 3 x 225. Draw the set models for the multiplier and the solution model.

What is the solution? Explain your solution in terms of expanded form.

_____

_____

_____

_____

_____

_____

_____

_____

_____

_____

7. Use bricks to model 5 x 125. Draw the set models for the multiplier and the solution model.

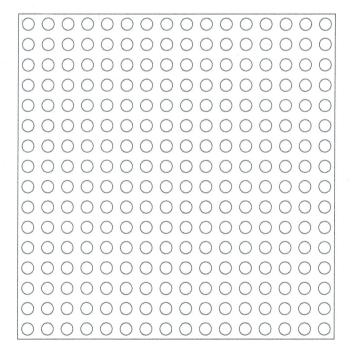

What is the solution? Explain your solution in terms of expanded form.

_____

_____

_____

Assessment:

1. What does expanded form show in a solution?

_____

_____

_____

2. What does the set multiplier show in a problem?

_____

_____

_____

3. Model the multiplication steps for 3 x 225 and explain all the steps.

_____

_____

_____

MULTIPLYING TWO-DIGIT NUMBERS BY TWO-DIGIT NUMBERS

1. With bricks, model 12 x 14 using an array model. Create the rectangle by placing bricks into the model horizontally and vertically to close the rectangle. Leave the ones bricks outside the rectangle. Draw this step in the base plate diagram on the next page.

What does this rectangle represent in the problem?_____

Label it on the diagram. (Remember to label all parts of your model throughout the process.)

With 1x4 bricks, fill in across the top of the rectangle (horizontally). What do these bricks represent? _____

Draw this step and label it on the diagram.

With 1x2 bricks, fill in the right side of the rectangle up to the top (vertically). What do these bricks represent in the problem? _____

Draw this step and label it on the diagram.

With 1x1 bricks, fill in the top corner of the rectangle. This will complete the model. These bricks represent the number _____.

Draw this step and label it on the diagram.

The model shows the expanded form of the multiplication problem 12 x 14.

Write each part of the model: _____ + _____ + _____ + _____ = _____

2. With bricks, model 16 x 11 using an array model. Create the rectangle by placing bricks into the model horizontally and vertically to close the rectangle. Leave the ones bricks outside the rectangle. Draw this step on the base plate diagram on the next page.

What does this rectangle represent in the problem?_____

Label it on the diagram.

What bricks do you need to use to fill the top of the hundreds rectangle (horizontally)?
_____

How many bricks do you need to use to fill the top? _____

Draw this step on the diagram and label it.

What bricks do you need to use to fill the right side of the hundreds rectangle (vertically)? _____

How many bricks do you need to use to fill the side? _____

Draw this step on the diagram and label it.

How many 1x1 bricks do you need to use to show the total ones? _____

Draw this step on the diagram and label it.

The model shows the expanded form of the multiplication problem 16 x 11.

Write each part of the model: _____ + _____ + _____ + _____ = _____

3. Using the array method, model all the steps to solve 13 x 12. Draw your model and label each part. Explain your thinking. Include the expanded form of the number in your explanation.

4. Using the array method, model all the steps to solve 11 x 11. Draw your model and label each part. Explain your thinking. Include the expanded form of the number in your explanation.

More problems for practice:

5. Using the array method, model all the steps to solve 14 x 13. Draw your model and label each part. Explain your thinking. Include the expanded form of the number in your explanation.

_____

_____

_____

_____

_____

_____

_____

_____

_____

_____

_____

_____

_____

_____

Assessment:

1. How does an array model show the expanded form of a multiplication problem?

_____

_____

_____

2. Model the problem 12 x 12. Show and label all parts of the problem and explain your thinking.